ROBERT BURNS ◆ FARMER

GAVIN SPROTT

N·M·S

NATIONAL MUSEUMS OF SCOTLAND

right: Mossgeil. DO Hill, *Land o' Burns,*
Edinburgh University Library

front cover: Robert Burns during his last year at
Mossgeil. This engraving by John Beugo is said to
be the closest likeness. Scottish National Portrait
Gallery

inside front cover: The farm of Shanter lay near
the Maidens rocks in Carrick, and the farmer
Douglas Graham was probably the Tam of the tale.
Here he is riding the small sturdy *pownie* common
in Ayrshire at the time, with no saddle or stirrups,
and only a simple home-made bridle. These animals
were used for personal transport, carting and
ploughing and earlier as pack horses. In Burns' day
there was a considerable import of horses from
Ireland into the west of Scotland. James Howe,
National Gallery of Scotland, SEA, NMS

title page: Leading sheaves on a hill farm. *Scotland
Illustrated,* SEA, NMS

**Published by the National Museums of Scotland, Chambers Street,
Edinburgh EH2 1JF.**

Photography: Ian Larner, Doreen Moyes
Picture research: Lorna Ewan, Alison Cromarty,
Elizabeth Robertson
Produced by the Publications Office of the National Museums
of Scotland:
   Editor: Jenni Calder
   Series design: Patricia Macdonald
   Design: Elizabeth Robertson assisted by Alison Cromarty
   Maps: E Helen Jackson
Typeset in Baskerville by Key Topic, Glasgow
Printed by Alna Press, Broxburn, West Lothian
© Trustees of the National Museums of Scotland 1990

ISBN 0 948636 18 1

Preface

Few readers of Burns' poetry appreciate the rigours of the farming life to which he was brought up. Farming in Ayrshire in the latter part of the eighteenth century involved an arduous and demanding routine. The domestic architecture, the equipment, the animals, the techniques, the whole way of life, are all present in the most detailed and intimate way in Burns' poetry. Reading Gavin Sprott's account of all this, with its precise detail and its careful pointing to the reflection of the farming life in the poetry, we are given a new awareness of the context and meaning of what Burns wrote.

We are also given a new realization of the almost miraculous nature of Burns' achievement. Amid the hard grind of the agricultural labour described here Burns was able to acquire a rich general education and to engage in a colourful social life. And, most of all, he was able to write his poems. We marvel at the way in which sheer genius broke through the harsh conditions he endured. Well-read, sensitive, perceptive, able to relish the sensitivities of the very animals he worked with or encountered, Burns was also keenly aware of every physical detail of his working environment, and the language of much of his poetry bears testimony to this.

We have learned enough about Burns' education to repudiate Henry Mackenzie's description of him as a 'Heaven-taught ploughman'. But we must not forget that he was a working farmer for most of his life, and that he acquired his book learning by sheer determination in the midst of arduous physical toil. Though he could imitate Pope and Shenstone and make gestures towards the literati of Edinburgh who patronized him without understanding him, his finest poetry is resonant with the language of his own class and occupation; this both anchors it in the concrete realities of his own life and time and helps to give it a richer human significance.

David Daiches

ROBERT BURNS: FARMER

A FOREIGN COUNTRY

The poetry of Robert Burns has such enduring freshness that it is easy to forget that the world has grown two centuries older since it was written. Between then and now, each generation has perceived Burns anew, after its own fashion. He has been the revolutionary, the patriot, the freethinker, the hero of romance, the ploughman-poet, the libertine and womanizer, the drinker and reveller, the folk hero. But what was he?

Robert Burns was a farmer. It was thus that he supported himself through most of his short life. The world from which he grew was that of the farms and villages of Lowland Scotland. To stand and sniff the weather as the day dawns, to watch the crops braird, to sort the beasts and handle the sheep, to work without rest until the harvest is safely in, these are the things he had in common with other farmers from the earliest times to those who work the land today.

Yet Burns farmed at a particular time and place. As a laddie he helped at his father's smallholding at Alloway (until 1766), and while still a youth, he was graduating to a man's work at Mount Oliphant, also near Ayr (1766—1777). At Lochlie near Tarbolton, he shared the work fully with his father William and brother Gilbert (1777—1784). On his father's death he moved to Mossgeil, near Mauchline, where he shared the tenancy with Gilbert (1784—1788). When Burns moved to Ellisland (1788) on the banks of the Nith, north of Dumfries, he was sole tenant for the first and only time.

If those different places provided a variety of experience, the times in which Burns lived provided even more, for the whole fabric of life was changing. Part of an island on the north west of Europe, Scotland had a long tradition of maintaining a distinct and lively culture on the short rations of subsistence farming and limited trade. But Scotland now became a full participant in the development of industry and a great maritime empire, and that changed the old ways. In popular imagination this figures as factories,

above: The *Hairst Rig,* looking across to Arran from Ayrshire. Showing a changing countryside: the old castle on the headland abandoned for a new house, old forest by the river, new thorn hedges enclosing the parks, grass integrated with the arable, and what were then modern houses and steadings.
P Hately Waddell, *Life and Works of Robert Burns*

opposite: The Carron Ironworks near Falkirk were started in 1760. Burns compared them to Hell. In fact, Carron helped to revolutionize agricultural technology, turning out iron castings such as the curved mould-board to James Small's improved swing plough.   Falkirk Museums

mines, smelters and urbanization. However, one of the pillars on which this new world rested was a stable and adequate food supply, and to provide this the countryside underwent a rural form of industrialization.

Burns not only saw this take root in Ayrshire and Dumfriesshire in his own lifetime, but he was caught up in it. His heart lay with the old close and familiar bonds of the unindustrialized countryside, but his head was with the new improved farming, for he knew that was the only way to make an even half-decent living.

Robert Burns was thus a man caught between two worlds. His poetic genius was informed by an equally remarkable intelligence. There were few people he did not communicate with easily, and little he did not notice. The tensions and contrasts thrown up by change are one of the ingredients

in the poems, letters and journals. However, what was familiar to people who knew Burns, and the many more who first read his poems, is not familiar to us. The freshness of Burns' poetry can be deceptive, for as in Hartley's now famous phrase, 'the past is like a foreign country'.

Our insight into the past is always limited, because that is not where we are. Nevertheless, there are various ways by which we can get at least some idea of what the past was like, and one of the most interesting is to explore the material culture — that is the physical fabric of people's lives. This includes the work they did, the land they farmed, and the houses they lived in. The study of the surviving objects and related evidence is an important function of the National Museums of Scotland. This can sometimes reveal the reasoning behind habits and ways of doing that seem irrational or foolish to later generations. With this evidence we can furnish our imagination with physical fact, and, even if only in part, recreate the everyday framework which shaped people's lives and generated their experience.

There are many references to this material world in the poems. They are, however, the few visible pinnacles of a reef submerged in history. Here we chart the outlines of the country life familiar not only to Robert Burns, but also to the first people who delighted in his genius.

Farming before the Agricultural Revolution

At the time of the Union of 1707 the country was farmed by a system that had matured in the Middle Ages. It was suited to a population scattered in isolated groups, and was heavily dependent on the natural resources of timber, peat and *naitur gerss* or self regenerating grazing. Markets were local affairs and although there were well established patterns of trade and communications, they provided only a minor part of the economic turn over. Produce was mostly for local consumption. Far from being the simple life, it was one of complication that we would probably find burdensome and baffling. People had to rely on their ingenuity to find resources, from the materials to build and heat their houses to means of coping with famine and disease. The arrangements by which communities pooled their labour and shared their resources demanded a sharp memory and a good head for calculation if friction were to be avoided.

There were certain general features common to the whole country. The ground, both Highland and Lowland, tended to be divided into arable and grazing, and while families would have particular areas to cultivate, the

Although this image suggests a rural idyll, children were set to work as soon as possible, particularly in herding animals on the open grazings.   George Harvey, engraved by Lumb Stocks, *Songs of Burns*

A *soutar's* or shoemaker's shop. Shoes or boots were either a necessity for rough work or a luxury for Sundays.　Walter Geikie, National Gallery of Scotland

grazing was held in common by a whole community. With livestock, the emphasis was on living off their products such as milk and wool, and their strength in the plough or for transport. To consume animals direct as meat was only a last resort, when there was not sufficient winter feed to sustain them, or as a luxury, such as at a wedding feast.

The old-style farming is often referred to as *run rig* after the general character of land tenure. Within the system there was some variety. In the Highlands and Islands the emphasis was more pastoral. In the Lowlands arable farming was more developed, and there the burghs provided at least some outlet for trade and centres of craft skills. The market afforded by the Capital stimulated a modest commercialization of farming round the Firth of Forth, so that when improvement came there was already some development to build on.

In years of plenty the old-style farming could afford a sufficient if simple, living, but it was unstable. There was a severe imbalance in the feed available for livestock between summer and winter. The increase of population was forcing more ground under the plough and diminishing the reserves of natural grazing. The arable was also narrowly based on grain and legumes. A bad season, and it was hunger. A really bad season, and it was famine.

The Ayrshire into which Robert Burns was born in 1759 was still farmed according to the old ways. However, these had themselves been changing. The old shared tenancies, where several families rented a farm, were dwindling. This did not affect the sharing of grazing. The neighbouring arrangements, whereby people helped one another with ploughing and harvesting, were still common. Horses had replaced cattle in the plough, effecting a change in the balance of livestock. There was not a widespread race of sub-tenants working to the tenants as in other parts of the Lowlands, but rather an even spread of small farmers, working with their sons and daughters, and unmarried servants who were part of their households. The cottars who lived in separate houses were not as numerous as on the east coast, and were often semi-destitute, having fallen on hard times.

LAND USE

The arable land was divided into *croft* land which lay by the farmstead, and the *outfield*. The croft land was under almost constant crop, with a simple rotation of perhaps three seasons of oats, one of *bigg*, an old form of barley, and sometimes a crop of pease, and increasingly at this time, one year of

fallow. The outfield was cropped with oats for three or four years in succession, until the return did not repay the labour, and the exhausted ground was left as grazing to recover for five or six years. The croft land got most of the manure, the outfield very little. Beyond the outfield lay the permanent grazing, in Ayrshire often just *the muir*, and generally known as *commonty*. On this surrounding farms ran their beasts according to their *souming* — the proportion of animals their arable holding entitled each farmer to.

In appearance, this unimproved landscape would be startlingly unfamiliar to the modern eye. Dykes were limited to those by the houses round the kailyard, and the cornyard containing the stacks, to protect them from the stock when it was inby during the winter, and if it were needed, the *heid dyke* that marked off the arable from the grazing.

In the days before improvement, the old *rigs* stretch out in front of houses and steadings on the edge of Hamilton. The moors and hills beyond were shared grazing.   John Slezer, *Theatrum Scotiae*

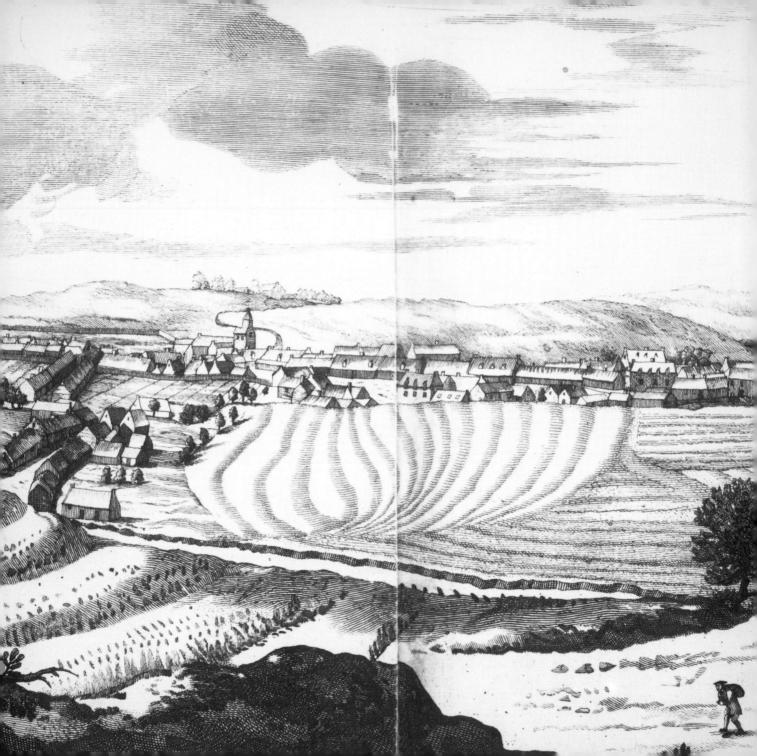

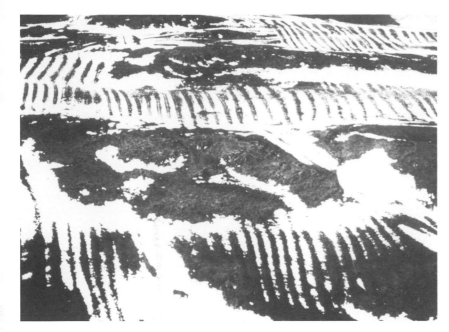

Old *rigs* in winter, viewed from above. Frost or snow often picks out the lie of rigs, which run down the slope to help surface drainage. Improvers kept the rigs, but straightened them for ease of working. The field surfaces were levelled from the 1830s onwards, when sub-soil ploughing made the field drains put in by the earlier improvers really effective.   Scottish Development Department

opposite: The pre-improvement *rigs* can often be distinguished by their curved shape. This was produced by the long team of oxen turning at the headland while the plough was still in the ground. Oxen had been given up in Burns' Ayrshire, but the rigs remained.   John Slezer, *Theatrum Scotiae*

The ploughed ground lay in *rigs* or ridges — strips of land that often snaked in shallow curves, up to thirty feet broad in the middle and tapering at the ends. The *croun* or middle of the rig could be three feet or more higher than the space that lay in between. The rigs were a way of effecting surface drainage. The spaces between the rigs ranged from a simple shallow ditch to *bauks* or unploughed rigs onto which stones from the rigs were tumbled. At best they provided little reserves of winter feed, but they were commonly patches of whin, broom and brambles.

However, even these weeds had their uses — the whin wood for the fire, and the tips could be bruised for fodder, and the broom used for thatch. Thistles grew everywhere, and they were gathered as fodder for the horses.

Water could become trapped between the rigs, and the drift of undrained moisture could render the lower margins of the fields unworkable, and there pockets of grass would be allowed to grow. From these and the hauchs along the burn sides people could gather bog hay. It was often little better than straw, but a valuable addition to the meagre supply of winter feed. More valuable was the *pease strae*.

Timber was more scarce than now, although something of the old forests survived down the banks of the main rivers. Planting trees seems to have

Detail of the bridge over the Allan Water at Dunblane. Bridges, important links in communication, were sometimes erected by wealthy public benefactors. The alternative was a ford. The river bed was sometimes *shod* or lined to give an even footing, or a smooth passage for wheeled vehicles.   John Slezer, *Theatrum Scotiae*

Detail of a slide car.   John Slezer, *Theatrum Scotiae*

preceded the main onset of improvement. The bonnet lairds fostered stands of timber round their steadings, and likewise the policies of the estates were graced with timber. The Ayrshire landscape does not seem to have been so denuded of trees as other parts of the Lowlands.

TRACKS AND LOADS

Transport was mainly by packhorse. The load was slung evenly on both sides from a wooden pack saddle which had lugs for attachment. Sheaves or hay could be loaded on *cadges* or side racks, and peats loaded in *creels* or large baskets. Manure from the byre or midden could be carried out to the fields by packhorse, and grain for the mill would go in sacks made of *pack harn* or coarse linen.

The alternative to the packhorse was the cart with *tumbler* wheels — discs made up in solid timber on a fixed axle. There were various forms of *cars* and *slypes* or sledges, some with long shafts of which only the rear ends trailed on the ground, and others with runners which were useful for low-loading stones and heavy objects.

Made-up roads did not exist, but rather routes or tracks. The open nature of the countryside did not confine transport in the way that the later enclosures did. One of the principal items of trade — cattle — moved themselves, hence the importance of droving. More important than roads were bridges, as rivers were the big obstacle to livestock movement. Although they were far and few between, the bridges such as those at Ayr and Alloway were very important.

above: The combination of horse or pony and creel was very versatile, and was once common to Lowlands and Highlands. Although painted against a Highland background, the pony with the peat creels probably contains Lowland blood.   William Shiels. NMS

left: Detail of leading a crop.   John Slezer, *Theatrum Scotiae*

HOUSES AND STEADINGS

The houses of the farmers were in quite stark contrast to what came later. The improvers sometimes described them as hovels, and compared with the improved housing some of them doubtless were. Nevertheless, the old pre-improvement houses had their own logic. They were integrated into the organic cycle of farming. When its day was done, the straw and divots of the thatch, enriched by soot from the fire, and the fibres broken down by the heat from the house below warming the sodden mass, could be recycled as manure. The ash from the peat fire was added to the litter in the byre, providing valuable potash in the manure. Because the roof was carried on *couples* or crucks that sprang from near the ground, the turf component of the walls could be stripped out independently and likewise be recycled, to rot down with muck, weeds, the scourings of *stanks* or ditches and even dead dogs and cats. It was all part of that formidable mass of putrefaction, the *midden*, the effluent of which fed the *dub* that lay not far from every house door and was the resort of the ducks.

Inside, the houses of the old-style Ayrshire farmers followed a basic pattern common to most of the country — a long-house which contained dwelling, byre, stable and barn, all joined in a straight row. The dwelling and byre, which sometimes included the stable, were the central unit. This was divided by a central passage, the *throwe-gang*, that connected the front and back doors. The front door gave out on to the midden, and through it went people and beast alike. The back door gave on to the *kailyaird*. The *cornyaird*, which contained the stacks of the grain crop, lay behind the barn.

An east coast example of a house with steadings, an enclosed cornyard containing stacks, and a kailyard. This picture shows late seventeenth-century changes. A substantial house of lime mortar construction, with a *hingin lum* at one end and a fire with a stone flue at the other, stands beside the older long house.   John Slezer, *Theatrum Scotiae*

The other buildings were variations on this pattern. The cottar's dwelling was a basic house and byre with no developed throwe-gang, but a simple *hallan* or partition dividing the people from perhaps a single cow and calf. In Burns' youth some of these cottar-houses would have had no *lum* or chimney, but simply a hole at the *riggin* or roof ridge for the escaping smoke. Besides the barn and stabling, the more developed farmstead might have a sub-division of the living space. This was what distinguished the bigger tenants and bonnet lairds from the rest. The throwe-gang gave access to the *in-seat* or main kitchen and living area, and *ben* or beyond that was the *spence*, the private apartment of the *guidman* and *guidwife*, and where they and their younger children slept. The expression of a house as being a 'but and a ben' — with an outer and inner part — has come to mean a modest dwelling. In mid-eighteenth-century Ayrshire, it was a sign of prosperity.

The in-seat served many purposes. It provided sleeping space for the servant lasses and older children, and if they did not sleep above the stable, for the unmarried men. This was made possible by box-beds, virtually small rooms within rooms with sliding or hinged doors on the front. These beds could form part of the partitions that subdivided the house, and were in the

The mid-floor hearth depended on peat for fuel. Although the *reik* permeated everything, it hung in the upper space, and did not hurt the eyes as wood or coal smoke does. Hearth type affected diet. The open fire was suited to baking on a girdle, grilling on a brander, or boiling in a pot. In some parts of the Lowlands, *breid* still refers to oatcakes, the product of open-hearth cooking. Wheaten loaf was a luxury in Burns' day, and only spread among the farming population when they acquired the coal-fired ranges with ovens of the type made at Carron.   Anon, SEA, NMS

This picture of a house at Alloway shows many of
the changes Burns would have seen. A fire of coals
roars in a stone-built gable hearth and the room
has a ceiling. The dresser, *aumry,* box-bed and two
tables spell comfort, and the clock, pewter and
crockery are small luxuries. The ample supply
of drink on the clay floor suggests imminent
company.   Anon, SEA, NMS

early eighteenth century still something of a novelty in the countryside.
Imported *dails* or timber cut into deals or planks were becoming available,
and this was just one sign of awakening trade. This new supply did much to
transform the plenishing of the existing housing.

Although the roof space was normally open, in the bigger houses part
could be divided off by flooring between the ties of the couples, and this could
provide additional storage or sleeping space. Valuables might be stored in
an *aumry* or press in the spence, and there also might be the *girnal* storing
oatmeal.

In the simple cottar-houses the plenishing would have been scanty. By the
time of Burns' death, they are described as having box beds. It is possible
that when Burns was a boy, some of the cottars would have still slept according to an older pattern, on beds of heather and brushwood on the floor
round the fire. Other plenishings might have included a simple *settle* or
bench and *creepies* and *buffits* — stools, and a *kist* or chest for valuables. Much
of the storage would have been in creels and close-woven straw-work
baskets.

An important item in the farmhouse living space was the dresser. This was
where food was prepared, and the dishes of cream put to settle before being

opposite: Girdle, pot, *brander, bannock spade, cruisie,*
and *saut backet,* would all be near or at the hearth.
The wooden *caup* was for eating out of, often with
a horn spoon.   NMS

Woman with a child on her lap.   Walter Geikie,
National Gallery of Scotland

made into butter. In a better-off household there might be a table pushed up next to the small window. On the wall there was commonly a rack to hold the grander pewter or simpler earthenware plates or wooden trenchers. Crammed into various corners and over the couples was a variety of vital objects: the *kirn* or butter churn, *chissits* or cheese vats, *boynes* or tubs, *luggies* or pails, and the water *stoup*, all these the products of the cooper's skill. In a corner was the *lit-pot* for stale urine used for fixing dyes, and in any well-found household, a spinning wheel, which would be in constant use.

The floor of the spence might be of timber, but that of the living space was a mixture of clay and a fine aggregate such as smiddy ash. When well trodden and repeatedly swept out with sharp sand, it would take on a hard glaze.

AT THE FIRE SIDE

Dominating all this was the fire. In the Ayrshire guidman's house, this was a square area near the internal gable, and over it, supported on posts, was the lum, to funnel the smoke upwards. It was thus possible to get round the fire on three, and sometimes four sides, two of which would have *settles* or *forms*, that is benches with a back rail. In a full house these could also serve for sleeping. The gable end of the fire was the *ingle-neuk*, the cosiest corner. Being the driest, it was also the place of the *saut-backet*, or salt holder. Before the fire might be the guidman's easy chair, his throne as head of the household. Even as Burns was growing up, much of this was disappearing, and the distinctive *reek* of the peats changing for the acrid smoke of coals.

Generally throughout Lowland Scotland this living part of the house was the *fairmer's haa*. Not so easy to imagine now, it was a semi-public place. There children and servants dressed and undressed, there the household ate, held communal worship, and in the light of the fire, the cruisie lamp or tallow candle, whiled away the evenings in mutual entertainment, but never wholly idle:

> On Fasten-een we had a rockin
> To ca' the crack and weave our stockin;
> And there was muckle fun and jokin,
>   Ye need na doubt;
> At length we had a hearty yokin,
>   At sang about.

Epistle to J Lapraik

Shrove Tuesday;
spinning party

enjoy a good conversation

spell of work

Although illustrating a cottar's house, this shows a scene of reasonable comfort. The man in his chair dominates the household. Behind him a screen keeps draughts from the fireside, and above it two fowls are perched. The lack of natural light and the crowding in a small space are characteristic. David Allan, *Cotter's Saturday Night*, National Gallery of Scotland

The Gaberlunzie. For the destitute to beg was no disgrace, and the licensing of beggars was the crude provision of a social service. However, the kirk sessions were always waging war against *sorners* or young able-bodied people who they thought should be working. Homelessness was another problem, and in Burns' day and long after, there was a constantly drifting population seeking shelter and a little kindness in return for odd jobs. Walter Geikie, National Gallery of Scotland

Here the *gaberlunzie* or respectable beggar, or the *chapman* with his pack, could get a bowl of brose or kail, and a night's lodging in the barn in return for odd jobs and perhaps a tale or two. Neighbours would come and go at will. Privacy was a luxury. It was to the quiet places outside that young couples resorted, such as the banks of the Doon, or 'Sweet Afton's Water', or the corn rigs to which young Robert 'held awa to Annie'. In that society to be a solitary tippler was painful eccentricity: to engage in the uproar of cheerful social dissipation was quite normal. But like the great Henryson of old, in his imagination Burns described the pleasures of retreat into tranquil privacy:

> And when the day had clos'd his e',
> Far i' the west,
> Ben i' the spence, right pensivelie,
> I gaed to rest.
>
> *The Vision*

above: Whiling away the evening to the notes of the *stock 'n horn,* the simplest form of pipe, by then uncommon. The curtain conceals a bed, a *cruisie* lamp burns at the lintel of the *lum,* and the salt box hangs at the back of the fire. The older man drinks from an old-fashioned wooden *coggie.*   Alexander Carse, *Evening in a Scots Cottage,* National Gallery of Scotland, SEA, NMS

right: Courting was the principal enjoyment of young people, and the evidence suggests that they were discreet rather than inhibited. The fun and superstition that marked Hallowe'en and the New Year were often about who would marry whom.   George Gilfillan, *The National Burns*

The way Burns gave expression to the personal in what was an intensely communal culture must have been a delight at the time.

> Then, tho' I drudge thro' dub an' mire   *puddle*
>   At pleugh or cart,
> My Muse, tho' hamely in attire,
>   May touch the heart.
>   *Epistle to J Lapraik*

Alloway under snow. Few poets can equal Burns' vivid evocations of winter – bones crazed with the cold, snow drifting through ill-fitting doors, folk huddled over their fires.   Glasgow Museums and Art Galleries

However, although the best room at Mossgeil was still called *the spence*, it was not the *ben* room of the older Ayrshire houses, but part of a modern house built only a few years before by Gavin Hamilton who had sublet the farm. Likewise the houses at Lochlie and Mount Oliphant were then fairly

Unites in common recreation;
Love blinks, Wit slaps, an' social Mirth
Forgets there's care upon the earth.

That merry day the year begins,
They bar the the door on frosty win's;
The nappy reeks, wi' mantling ream,
An' sheds a heart-inspiring steam;
The luntin pipe, the sneeshin mill,
Are handed round wi' right good-will;
The cantie auld folk crackin crouse,
The young anes rantin thro' the house—
My heart has been sae fain to see them
That I for joy hae barket wi' them. —

Still it's owre true that ye hae said,
Sic game is now owre aften play'd;

modern. *The auld clay biggin*, Robert's birth place at Alloway, followed the old pattern of aligning the stable and byre with the house, but the walls were of the more modern clay mortar, and the roof was supported on modern wall-head couples, not the massive old couples springing from near the ground. The kitchen fire was set into a solid stone gable hearth with a flue in the thickness of the masonry, suitable for burning coal.

The house had been built by William Burnes, Robert's father. He was a Kincardineshire man who had also travelled and worked in Midlothian. There he perhaps picked up a newer fashion of building, but without the knowledge, for one of the gables collapsed when Robert was a child. Although Burns was familiar with the old houses of the Ayrshire countryside, he did not live in them. Likewise he understood the character of the old-style farming, and although he inherited many of the traditional skills, he was not an old-style farmer as his father's father was in Clochnahill in the north of the Mearns, nor his mother's father who was in Craigenton, by Kirkoswald. His heart lay with the old world that was passing, and his head with the new, for that was the only way to make a living.

Change

When Burns died in 1796 the Ayrshire of his birth had changed quite drastically. Within the compass of thirty-seven years, the pattern of land use, the balance of livestock, transport, housing and the population itself were all different. By the end of this period the change in farming technology was becoming evident. This was the Agricultural Revolution as it is known to history, or *improvement* as people then called it.

There were several factors which drove forward this astonishing change. The old-style farming may have been changing in itself, but the system could not adapt sufficiently to accommodate the increasing population and the shortfall in natural resources. Burns' generation did not have the experience of actual famine, but he would have talked to people who had. Even as the improved farming was getting under way there were close shaves. The summer of 1782 had been cold and wet throughout Scotland. When a fearsome frost struck on 5 October much of the crop was still green and standing. The Burns family was then in Lochlie, which was near the southern limit of the snow which fell before what had survived the frost had been gathered in. 1784 and 1785 were also bad years. For Robert and his brother Gilbert, then in Mossgeil, this was compounded by poor seed in the first year. The weather

The road to the Braes of Glenbervie. Here the *Burnes* or *Burness* family had tenanted various farms, including Bogjurgan, Brawliemuir and Clochnahill. Although brought up an Ayrshire man, Burns was very aware of his north-east ancestry, and visited his cousins and his father's native district in 1787.   SEA, NMS

opposite: Lines from *The Twa Dogs* in Burns' hand. Here Luath describes the cottar-folks' ability to enjoy themselves despite their poverty. Kilmarnock and Loudoun District Council, Dick Institute, Kilmarnock

above: Patrick Miller of Dalswinton was a man of great enthusiasms, even buying Dalswinton, in which Ellisland lay, without seeing it. His ideas were often ahead of practical reality. He pushed forward improvements on his estate but they only bore fruit after Burns was gone.   Sir George Chalmers, National Portrait Gallery, London

right: Burns' letter to Miller of Dalswinton, discussing the *tack* of Ellisland. He took it for a rent of £50 a year, rising to £70 after three years. Miller put up £300 for building a house and steadings, and what remained would go on improvements, which would include dyking, hedging, draining and liming.   Trustees of the National Library of Scotland

Sir,

I send you Mr gordon's scroll, and another which a professional man, a friend of mine, has done today. — This last is, I think, more distinctly what we have mutually agreed on; particularly the 300 £. —— According to your idea, I have mentioned the applying the surplus, if any be, to the improvement of the land; and as I told you, I wish to keep 50 £. of the first rec? monies, to be the latest accounted for, in case my stock be rather scanty. — There is some fishing rights the present tenant possesses; if you intend that I should enjoy the same, it will be best, I suppose, to mention it in the tack; if you do not understand that I am to have that priviledge, 'tis but a trifling matter, and I do n't much care. — If this scroll meets your approbation, I shall wait on Mr Gordon to get the tack extended, so soon as you return me the papers. ——

I have the honor to be ever, your highly oblidged and most respectful humble serv?

Rob! BURNS

S! Ja! sq?
Sunday even:

26

was thus the cruellest tyrant of all, and the Burns family was only one among many which it nearly ruined.

During the harvest of 1785 Burns wrote what was every farmer's prayer:

> May Boreas never thresh your rigs,
> Nor kick your rickles aff their legs,   small stacks
> Sendin' the stuff o'er muirs an' haggs   marshes
> Like drivin' wrack;   weed
> But may the tapmast grain that wags
> Come to the sack.
>
> *Third Epistle to J Lapraik*

John, 4th Earl of Loudon, on whose ground Mossgeil lay, was one of the biggest improvers in Ayrshire. He took an active interest in farming, promoted the building of new roads, and was the first to bring turnip cultivation to Ayrshire. He chose his tenants personally and got to know them.   Allan Ramsay, Scottish National Portrait Gallery

Improved farming was more flexible and reliable than the old ways. It secured winter feed for the stock by producing fodder crops. For the new sown grasses to thrive, the soil had to be put into a much better condition. Through liming and draining the ground became sweeter, releasing the natural nutrients, and because it was less waterlogged, it became 'warmer'. The seed would not rot in the ground so readily and would *braird* or germinate more surely and earlier, effectively bringing forward the growing season.

The means of effecting improvement were also coming together — a combination of wealth, organization and the growth of markets. An important part of this framework was the estate. This unit of landholding was a little kingdom, with the *laird* and his lady as sovereigns. The laird let his land to his tenants in return for rent, and sometimes services. However, activity was subject to laws of contract, even if that was largely framed by the landowning interest. This covered the vital item of the *tack* or lease by which the farmer held his land of the laird. The tack was all important. It gave the farmer tenure, but also dictated the terms. If the laird wanted to change the way the land was farmed, the basics were written into the tack. The tenant was compelled to stick to it by law, or pay a heavy fine, or be turned out.

The Ayrshire lairds varied enormously in status and wealth. The *bonnet lairds* — small owner occupiers — had neither the capital nor inclination to change. There was a considerable body of gentry with small estates. Many of these were in debt, a burden passed from one generation to another, with some of the land tied up in *wadsets* or lease-lend arrangements. These people often made ends meet by turning to the law, the navy or the army as professions, and at first their interest in improving the land was limited both in financial capacity and interest. It was only in Burns' time that those

Peeling potatoes. When Burns was born, potatoes were hardly known as a field crop in Ayrshire, but during his lifetime they became common and a valuable addition to the diet. The potato had tremendous advantages. It yielded more food per acre than any other crop, and it was easy to plant, harvest and store.   Walter Geikie, National Gallery of Scotland, SEA, NMS

who had worked abroad began returning in sufficient numbers with wealth from the West Indies or India. People were also beginning to buy land out of the fruits of local business. Most of the farms rented by the Burns family were from those who had made money in some form or another. By that time there were also examples of improvement to follow. There was the progress already made in Berwickshire and the Lothians, and the first tentative patterns set up by the greater lairds in Ayrshire, in particular the Earls of Loudon and Eglinton, who had sufficient land and capital to take the plunge.

THEORY AND FACT

The foundation of improved farming was a change in the way the land was used, and from that all else followed. There was to be no more separate arable and grazing. By turns all ground would come under the plough, the grass be sown artificially, or alternative fodder crops of potatoes or turnips sown. In the event, turnips were to be a comparative rarity in Ayrshire even by the turn of the century. The mainstay of the fodder crop was rye grass and clover, and to a lesser extent, potatoes. There was a simple framework for this. Only one-third of the farm could be cultivated at one time. Thus any bit of ground would be in crop three years, produce hay in the fourth, and be in *lye* or lea for the remaining five, after which it would be broken up again.

The use of lime was known in Ayrshire in the previous century, and the first experiments with sown grass were on the Loudon estate in 1735. The man after whom the Ayrshire rotation was called, Alexander Fairlie of Fairlie, introduced it on his ground in 1744. However, such experiments were isolated examples. If there is a significant general starting point, it is the later 1760s, when William Burnes had just taken the tenancy of Mount Oliphant. In 1766 the first turnpike trust to start made-up road construction in Ayrshire was formed. The enclosure of farmland started in earnest, and Fairlie of Fairlie was appointed commissioner for the Eglinton estate, following the death of Earl Alexander in 1769.

These factors were interlinked. If the arable and grazing were to be integrated, the livestock would be grazing in fields adjacent to those that carried a crop, and they had to be controlled. The traditional means was by tethering the cattle to graze specific areas, or setting a *herd-callant* — a herd laddie — to mind them. This he could do with a stick, or at longer range by pelting straying animals with stones. However, the planting of thorn hedges in the

lower parts of Ayrshire, and the building of dykes eventually did away with this need.

This enclosing defined not only the fields but the roads. Until road-building advanced, with ditches, *cundies* or culverts, and *metal* or well-broken stones, the roads actually got worse before they got better. This communications revolution was the life-blood of farming improvement. The backbone of the new transport was the two-wheeled horse-drawn cart. It had a capacity of at least four or five times that of the old tumbler carts. The main uses were *leading* or bringing in crops, driving out dung, the movement of coals and lime, and getting produce to mill and market.

With his appointment as Eglinton factor, Fairlie now controlled sufficient land to make a widespread enforcement of his new rotation, and this set a new pace for enclosure and road-building. As matters progressed, it was found possible in some cases to modify his rotation from a third of the ground in cultivation to a quarter. Another obstacle to change receded, as the old fifty-seven year tacks began to run out and the ground was re-let on a new basis. There was also a significant connection with the Burns family. William Burnes first came to Ayrshire to work as a skilled gardener for Alexander Fairlie. Although in personal terms he was a dour old

above: A box cart at Alloway.   D O Hill, *Land o' Burns*

left: Ayrshire cows at a troch.   Walter Geikie,
National Gallery of Scotland

Looking over the Field of Bannockburn across Central Scotland. In 1787 Burns visited the sites connected with Wallace and Bruce. Burns' reverence for his hero stemmed from reading Blind Harry's *Wallace.* Together with the *Bible, Pilgrims Progress* and other religious works, Hamilton of Gilbertfield's version of *The Wallace* was read widely by lowland farming people. At the time lowland Scotland had one of the highest levels of literacy in Europe. Without that Burns could never have enjoyed the success he did among his own people.   DO Hill, *Land o' Burns,* Edinburgh University Library

conservative, William Burnes appears to have been a highly intelligent man, and he was wedded to the new methods.

However, between theory and practice there were significant gaps. At first improvement was an imported fashion that was imperfectly understood. Old snaking rigs were enclosed as they lay, and realigning them into the improved straight fifteen-foot rigs that became fairly standard was a problem. The lairds borrowed money to reshape the land and expected a miracle. Between them and the land were the tenants, and it was on them that improvement could bite most cruelly. In his time, Burns travelled widely — southern Argyll, the central Highlands, Buchan, the Mearns, the Lothians, Berwickshire and Dumfriesshire, his farmer's eye noting what he saw. His conclusion on the Ayrshire lairds was cutting. He wrote to his Kincardineshire uncle 'our landholders, full of ideas of farming gathered from the English and the Lothian and other soil in Scotland, make no allowance for the odds of the quality of land, and consequently stretch us much beyond in the event, we will be found able to pay'.

POOR TENANT-BODIES

Until Robert was seven, the Burns family lived at Alloway, on the seven-and-a-half-acre holding that William had feued. In 1765 William became a tenant farmer at Mount Oliphant, not far from Alloway, then about eighty acres (imperial). This was perhaps the hardest part of Robert's life. Robert's brother Gilbert recalled that the ground was so wretched that even thirty years after the Burns family had left and considerable sums had been spent

opposite: Mount Oliphant today. Careful drainage has changed what was wretched ground, but not until long after the Burns family had left.   SEA, NMS

Gilbert Burns. He outlived his elder brother by
thirty-one years. Gilbert remained at Mossgeil until
1798, spent two years at Dinning in Nithsdale, then
moved to East Lothian, where he spent the rest of
his life. In 1804 he became factor to Lady Blantyre.
Although a more consistent farmer than his
brother, he was also rescued at Mossgeil by
Robert's generosity with the proceeds of his
publications.   Trustees of the National Library
of Scotland

Burns' signatures, the second one that of a dying
man, July 1796. When Burns mentioned ill-health,
it was often in announcing his recovery. Although
Burns was erratic in attending to the farm work, he
did not spare himself once it had got his attention.
Ploughing, building dykes and threshing are only
some of the more strenuous tasks he did, hardly the
best medicine for a man who died at thirty-seven,
probably from heart disease.   Scottish Record
Office

on improvement, it would still fetch only £5 less rent than what his father had
paid for it. There was no spare money to hire any help, and even as a laddie
Robert was doing a man's work. The strain probably inflicted the damage
that led to his early death. And here, struggling to pay the rent, was the Burns
family:

> Poor tenant-bodies, scant o' cash,
> How they maun thole a factor's snash;   bear; abuse
> He'll stamp an' threaten, curse an' swear,
> He'll apprehend them, poind their gear,   impound
> While they maun stand, wi' aspect humble,
> An' hear it a', an' fear an' tremble!
>
> *The Twa Dogs*

This description was drawn from Burns' experience at Mount Oliphant,
which made a lasting impression. A naturally friendly man whose acquaint-
ance embraced the whole social spectrum, Burns never overcame his
distrust of factors, as the lackeys of oppression. Paradoxically, his brother

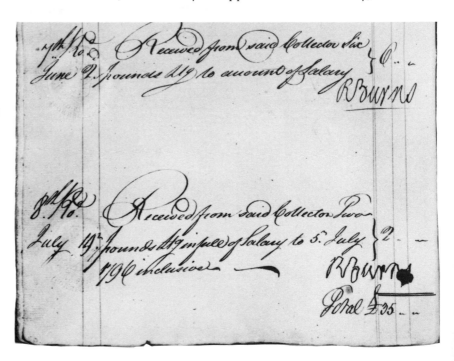

Gilbert, cautious, hardworking and decent, went on to become a factor in East Lothian.

They stuck Mount Oliphant for twelve years, then in 1777 moved to Lochlie, by Tarbolton, 130 acres. This enterprise was something of a financial gamble. Where Mount Oliphant had been twelve shillings an acre, Lochlie was to be fifteen shillings, or £1 if the laird, McLure of Shawwood, made certain extensive improvements — steadings, dyking, heavy liming, and draining the loch. There was some hired help as Willie Ronald was *gadsman* — the old name for the person who drove the horses when ploughing. More ground was broken up, probably to accommodate the new integrated rotations. As well as corn, barley and flax, they were growing

Collecting the rents. The factor was the executive head of an estate organization. He had to understand the legal framework of landholding and know the value of land and other natural resources. Before farming improvement, factors were often expert in juggling with debts, those of their master and of the less fortunate tenants. During Burns' time factors were evolving into professional land managers.   SEA, NMS

Lochlie today. NMS

Gavin Hamilton was a lawyer, who became factor to the Earl of Loudon, and as such was an exception in Burns' dislike of factors, for he was one of his surest friends and supporters. Hamilton had the tack of Mossgeil from the Earl of Loudon, and sub-leased it to Robert and Gilbert. Although Mossgeil was hard going due to circumstances beyond Hamilton's control, it was never the 'ruinous bargain' of Ellisland, the battleground of Lochlie, or the heartbreak of Mount Oliphant. Otto Leyde, National Gallery of Scotland

wheat, unusual for the area. It was when he was in Lochlie that Robert won a prize for flax, and went to Irvine briefly to learn more of the trade. In the event, it was the Burns family who carried out improvements, and William fell out with the laird over the finances. The matter went to court, and old Burnes won, but it was a sad victory, for he died soon after. As at Mount Oliphant, Lochlie was something of a struggle, but despite the court case the family left the place free of debt, which in these days was an achievement.

In 1784 the brothers moved to Mossgeil, by Mauchline, 118 acres, a sublet farm on the Loudon estate. Here the ground was high and exposed, with a cold wet bottom. It had probably been skinned for turf used for building in the not-too-distant past, and this degraded the soil by carrying off precious top soil. Despite the difficulties with the seasons, the seed and the ground, life at Mossgeil seems to have been easier. Besides the Burns family, there were farm servants, such as John Lambie, who led the horses when Robert turned up the mouse with the plough. John Blane and Willie Patrick are mentioned, and there were others:

> For men, I've three mischievous boys,
> Run de'ils for rantin' an' for noise;
> A gaudsman ane, a thrasher t'other,   driver
> Wee Davock hauds the nowt in fother.   keeps the cattle fed
>
> *The Inventory*

34

above: Mossgeil, just before the middle of last century. The gable of the barn, stable and cart shed just shows behind the house, and beyond that stacks stand newly roped and thatched in the cornyard. Although the house is still thatched, by this time the new byre and milkhouse in the foreground have been added.   J Kennedy, National Gallery of Scotland.

left: Mossgeil today.   NMS

No pictures exist of Jean Armour as a young woman. In her middle fifties she was still 'a very comely woman'. Although she had no passion for reading, she had a lovely singing voice and shared Robert's interest in the old ballads. Despite her husband's waywardness, she had the character to hold his affection. Of their three surviving children, not one became a farmer.   Scottish National Portrait Gallery, from the original at Glasgow Museums and Art Galleries

The kitchen at Mossgeil. The door on the left leads to a lobby where there was access to the small rooms in the loft where Robert slept and did much of his writing, and beyond that is the best room or *spence* as Burns still called it. Through the door on the right is the milkhouse.   Sir William Allan, National Gallery of Scotland, SEA, NMS

It was as 'Rab Mossgeil' that Burns' genius blazed into its full glory. He also applied himself seriously to making a living. 'I entered this farm with a resolution. "Come, go to it, I will be wise!" I read farming books. I calculated crops. I attended markets.' The poet and the farmer in him had to live together, but sometimes with curious results. It was common for two men to work together at the thrashing, wielding their flails turn about, but with Robert it was dangerous. He would work not to the time of his partner but to the lines or tunes running in his head. He would send carts away unempted, or lose his concentration when ploughing, his lips moving silently as he worked on a poem forming in his head. He relied on the servant lads and Gilbert to keep him right. This constant mental distraction exasperated Gilbert, and amused his servants, who recalled him as a kind and indulgent master.

Even at meal times Burns was in another world, spoon in one hand and book in the other. Later a visitor to Ellisland remembered that 'he lived all his days the inward if not the outward life of a poet. I thought I perceived in Burns' cheek the symptoms of an energy which had been pushed too far.'

above: Cattle at market.   Walter Geikie, National Gallery of Scotland

left: The Cotter's Saturday Night. Burns' poem had so caught the public imagination that it was a favourite for illustration. Here the family is shown round a table which has been drawn up in front of the fire. There is an *aumry* behind the group, and box-beds opposite the fire place. Even by the early nineteenth century, when this drawing was made, people had forgotten how cheerless the cottars' houses were.   Alexander Carse, National Gallery of Scotland

Despite this absorption by intense mental activity, Burns was a sociable man:

> We are na fou, we've nae that fou
>> But just a drappie in our e'e;
> The cock may craw, the day may daw,
>> And ay we'll taste the barley bree.
>
> *Willie brew'd a peck o' maut*

He enjoyed a good dram, but always in good company.

The farm at Ellisland, 170 acres on the banks of the Nith in Dumfriesshire, was meant to secure Burns a living on favourable terms that would enable him to pursue his interests. It was offered to him by Miller of Dalswinton. At the time Burns suspected that Miller's philanthropy was more idealistic than practical, but in 1788 he took the place, persuaded by John Tennant, one of his father's old friends. Once again he found himself trying to put heart into ground exhausted by constant cropping. He had to set to, clearing stones, ditching, hedging and dyking, and not least, supervising the construction of what was to be his and Jean Armour's new home. His life as a

below: Ellisland. The ground was stony when Burns came in 1788, and it is stony today.   Lorna Ewan

right: Ellisland, from the banks of the Nith. Burns superintended the building of the house and steadings.   DO Hill, *Land o' Burns,* Edinburgh University Library

above: Ellisland today, showing the house and part of the steadings. In 1921 the farm was bought by John Wilson, who formed a trust to look after it for the nation. It is open to visitors.   NMS

left: Burns started building the dykes at Ellisland. For him, dry stone dyking would not have been familiar. The older dykes in Ayrshire were of *fail* or sod, or a combination of fail and stone. The Ellisland dykes will have been rebuilt over the years from stones, some of which were originally gathered by the poet.   NMS

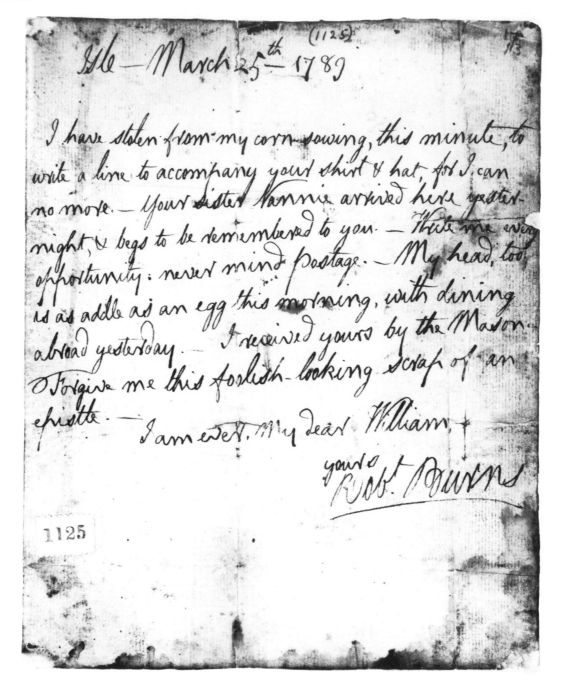

Ģle — March 25th 1789

I have stolen from my corn-sowing, this minute, to write a line to accompany your shirt & hat, for I can no more. — Your sister Nannie arrived here yester-night, & begs to be remembered to you. — Write me every opportunity: never mind postage. — My head, too, is as addle as an egg this morning, with dining abroad yesterday. — I received yours by the Mason. —

Forgive me this foolish-looking scrap of an epistle. — I am ever, My dear William,

yours
Robt Burns

farmer continued the seemingly endless treadmill of effecting improvements, with little of the rewards.

Of all the farms Burns was in Ellisland is now closest to the original, and the oldest buildings are in the same pattern of those once at Mount Oliphant, Lochlie and Mossgeil — the early improved arrangement of the byre and stabling as a separate building at right angles to the house. Ellisland was built by Alexander Crombie, and like most country buildings at that time, probably thatched.

Willie Clark, who worked at Ellisland over the winter of 1789—90, recalled a household that included two men and two women servants. Burns and his family now ate separately in the small parlour. There was a stock of about nine milk cows, some calves, four horses and several pet sheep. Clark

opposite: Robert Burns' letter to his youngest brother William, in whom he always showed a kindly interest. It describes two more distractions from his farm work.   Trustees of the National Library of Scotland

This anonymous drawing shows Mossgeil much as it was in Burns' day. Adjoining the house is the byre, barn, stable and cart shed. Everything is on a small scale. Burns usually ploughed with a four-horse *yoke* or team, like this one.   P Hately Waddell, *Life and Works of Robert Burns*

above: *Plaids* of the kind Burns wore were used in the lowlands into this century by shepherds, this one by William Arres. Besides weather protection, they were useful for carrying frail lambs. The broad Kilmarnock bonnets were a common part of working clothes into the first half of last century.   NMS

right: A man in working clothes.   Alexander Nasmyth, National Gallery of Scotland

Painted long after the event, this shows Burns in James Sibbald's circulating library, Parliament Square, Edinburgh, while the young Walter Scott looks on. Here Burns is dressed in the smart best clothes of a well-to-do farmer, a far cry from his working clothes. However, in the city his riding boots still pick him out as a countryman.   William Borthwick Johnstone, Edinburgh Booksellers' Association

repeated the report of a kind and familiar master. His image of him was not the well-to-do farmer, but a working man with his broad blue bonnet, drab old-fashioned long-tailed coat, corduroy breeks, dark blue stockings and *cootikins* or woollen leggings to keep the *gutters* or mud out of his shoes, and in the cold of winter, his black and white checked plaid.

As with Burns' other farming efforts, Ellisland turned out to be a hard bargain. It did not stop his writing, nor the considerable effort he put into the Scots Musical Museum, but it was still too much. Pushed to the limit of his considerable energy, and failing in health, Burns was lucky to get a post in the Excise service, and gave up farming in 1791.

THE WORK

Although improvement changed the patterns of land use, the technology was a mixture of the old and the new. Burns' time in Ayrshire was just before the general spread of the big Clydesdale horse from Lanarkshire. Ploughing was still with four horses, yoked in pairs, *fittie-land* and *fittie-fur*, the rear left and right, and *fore-wynd* and *fore-fur*, the front left and right. Burns also referred to them as *lan 'afore, lan 'ahin, fur-afore* and *fur-ahin.* The commands to the horses were common to southern Scotland, *hawp* — to the right, *wynd*

Although at Dunbar on the East Coast, this shows the ploughing arrangement with which Burns was familiar. One man guides the plough, the other drives the horses. The plough is an early improved version of the traditional Scots swing plough. It still has the heavy four-sided frame and possibly a flat mould board, requiring the draught power of two pair of horse. The heavier Clydesdale horse only spread at the time of Burns' death.   John Clerk of Eldin, National Gallery of Scotland

— to the left, *tsick tsick* — to go on, *pproo* — to stop. The bond between the farmer and his horses was a close one, an affection the old farmer had for his old mare, the *tocher* or dowry that bore his bride home:

> Thou was a noble fittie-lan'.
> As e'er in tug or tow was drawn!   leather; rope
> Aft thee an' I, in aught hours gaun,
>   On guid March-weather,
> Hae turn'd sax rood beside our han',   quarter acres; ourselves
>   For days thegither.
>
> *The Auld Farmers' New Year morning*
> *salutation to his Auld Mare, Maggie*

The four-horse yoke required one man to guide the plough, and someone to lead the horses. If the gap between the *coulter* or knife which cut away the side and the *sock* or ploughshare which cut away the bottom of the *fur* or furrow became choked with weeds and stones, it had to be cleared with a *pattle* or stick. This was the weapon with which John Lambie chased the mouse. In the event Burns stopped him, and finished *To a Mouse* by the following day.

The ploughs were lighter versions of the *auld Scotch ploo*, mostly timber, but with the *irons* — the sock and coulter, and the *muzzle* or hitch at the front of the beam, and the cladding of the breast and board in wrought iron.

Ploughing, harrowing and sowing were the main tasks early in the year. Although some of the old *lye* might be broken up in the back end of the previous year, the main work started about Candlemas, or 16 February by the old calendar to which many people still adhered for such reckoning, and carried on well into April. The seed was broadcast by hand. The sower had a sheet to hold the grain, usually over the left shoulder and cradled in the left arm, and cast with the right hand. Sowing was an unnatural motion that had to be learnt, hence Burns' comment:

> Forjesket sair, with weary legs,   exhausted
> Rattlin the corn out-owre the rigs
> *2nd Epistle to J Lapraik*

The seed was then harrowed in.

The drive to improve the condition of the ground made it easier to plough. This opened the road for the new light ploughs following the pattern developed by James Small in Berwickshire and adopted widely in the Lothians. The *mould board* — the curved plate which turns over the furrow — was made of cast iron. These were appearing in Ayrshire in the 1780s and were light enough of draught to be worked by two horses. Although Burns did not work with them, he would have seen them in use.

The preparation before ploughing meant carting out the muck from midden and byre, and at Mossgiel Burns mentions

> Three carts, an' two are feckly new;   almost
> *The Inventory*

Although drawn in the early 1820s, this shows the arrangement Burns would have seen over forty years before in East Lothian. The plough is Small's light improved version with a cast-iron mould board, and there is only one, not two pair of horse. The plough is guided and the horses driven by the same man. Only the harness is obviously nineteenth century. Here the *hind* or ploughman is sliding the plough round at the headland.

This plough was not just technically more efficient, but was possible because the improved state of the ground allowed an implement that demanded less strength of draught and was of lighter construction.   James Howe, National Gallery of Scotland

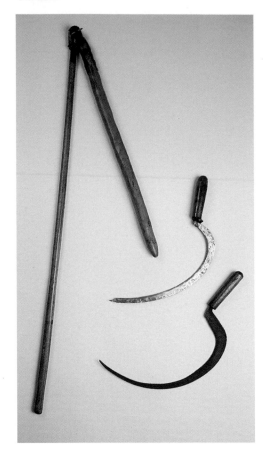

The older *heuk* had a serrated edge, and the straws were grasped and cut with a sawing motion. The faster smooth bladed *scythe-heuk* was spreading in Burns' day. The flail consists of the longer *hand-staff* or handle, and *souple* or beater.   NMS

Liming was also a major task that required carts to bring in the limestone and coals, and to spread the burnt lime. The Ayrshire farmers developed the skill of burning it in the fields in temporary sod-walled kilns. A fire of brushwood was started in the bottom, coal added, then broken limestone and more coal turn about, and the glowing mass left to cool for a day or two, covered over with sods. The resulting lime was then *slochened* or slaked, and spread as required.

The high point of the year's work was the *hairst* or harvest, and with the labour went the fun. The crop was sheared with the *heuk* or sickle, bound into sheaves, and stooked. In Ayrshire it was common to pair two young people to shear together. This custom was the start of many a romance, and with Robert it did not fail. At fifteen he was paired with Nellie Kilpatrick, the first record of his encounters with the opposite sex.

A far cry from the solitary work of today's combine, the harvest field was then busy with people, and Burns vividly recalled this scene of companionable labour:

> When first amang the yellow corn
> A man I reckon'd was;
> An' with the lave ilk merry morn   rest
> Could rank my rig and lass;
> Still shearing and clearing
> The tither stooked raw;
> With clavers and haivers   nonsensical talk
> Wearing the time awa'.
>
> *The Answer*

But if the weather was already breaking, the harvest could be a struggle:

> While at the stook the shearers cow'r   sheaves propped to dry
> To shun the bitter blaudin' show'r.   pelting
>
> *Epistle to the Rev John M'Math*

The labour of stooking might have to be redone more than once:

> But stooks are cowpet wi' the blast,   overturned
> An' now the sinn keeks in the west,
> Then I maun rin amang the rest.
>
> *Third Epistle to J Lapraik*

above: The *hairst* at Auchendinny, Midlothian, early nineteenth century. The harvest brought people together from every stage in life.   Anon, NMS

left: Harvesting with the *heuk* or sickle near Sanquhar. The use of this ancient tool was highly organized. A man and one or two women were teamed to straddle the *rig*. The man also made the *band* from lengths of straw on to which the cut crop was placed. Others came behind, closing the band and knotting it, completing the sheaf, which was then propped up against others to form a *stook*. DO Hill, *Land o' Burns,* Edinburgh University Library

The end of the harvest was the occasion for celebration. There were rituals in the cutting of the last sheaf-ful, which might be saved as an offering to the principal livestock on New Year's day, as the old farmer offered a *ripp* of corn to his old mare Maggie.

The barns would be clear of the previous year's crop, and not yet cluttered with the produce of the new, which would be standing in the cornyard behind, thatched and roped against the onset of winter. The stage would be set for the *kirn* or dance:

> Yestreen when to the trembling string
>   The dance gaed through the lighted ha',
> To thee my fancy took its wing,
>   I sat, but neither heard, nor saw:
> Tho' this was fair, and that was braw,
>   And yon the toast of a' the town,   farming township
> I sigh'd, and said amang them a',
>   'Ye are na Mary Morison'.
>
>   *Mary Morison*

The hard work of the *hairst* was usually sustained by better than average food and drink – perhaps the treat of wheaten loaf, or cold meat washed down with *sweet* or full cream milk, or ale. A common refreshment while working was water with a little oatmeal thrown in. DO Hill, *Land o' Burns,* Edinburgh University Library

opposite: An outdoor dance in Nithsdale. On the left is an inn, and on the right a barn, with a *wecht* or blind sieve lying above the eaves. The stackyard in the distance is full but the leaves are still on the trees, so the harvest is just in. The eighteenth century had seen a great revival of dancing as the rigours of puritan Calvinism were relaxed. Outside dancing was not uncommon, and here a fiddle and cello form the band. Burns had an impressive knowledge of native Scottish fiddle music, and met Neil Gow, the renowned Highland fiddler. DO Hill, *Land o' Burns,* Edinburgh University Library

There were various jobs that went on throughout a lot of the year, and thrashing the grain from the straw was one of the most tedious:

> The thresher's weary flingin-tree,   flail
> The lee-lang day had tir'd me;
>
>   *The Vision*

This was done in the barn with the flail. The barn was built with doors in opposite walls to create a through draught for winnowing. The doors could be left with the bottom half shut, or taken off the hinges and laid across, to keep the ducks and hens from raiding the precious grain as it was thrashed out. Between the doors was the thrashing floor, and a good through draught would carry the *caff* or chaff to one side, to be kept for filling bedding, and thus the *dichtin* or winnowing was done at the same time. However, the days of this drudgery were numbered. By 1788 James Meikle had developed the first effective thrashing machine, and this would undoubtedly have been an item of conversation when Burns dined with him at Duns during his Border tour.

An Ayrshire bull. *Stephens' Book of the Farm*

LIVESTOCK

The livestock was changing as much as the way the land was worked. The advantage of a stable source of winter feed was that breeding could be much more consistent. The control of the stock afforded by the new enclosures also made animal breeding easier. If there was a rule of thumb, it was to breed from the character of the second generation back.

The old native fine-wool sheep, not unlike the modern Shetland breed in general character, were giving way to the newer breeds of the Cheviot from the eastern Borders, and the Black-face, originally from the Pennines. As Burns noted of Poor Mailie:

> She was nae get o' moorlan tips,
> Wi' tauted ket, an' hairy hips;   matted fleece
> For her forbears were brought in ships,
> > Frae 'yont the Tweed:
> > *Poor Mailie's Elegy*

Burns had a soft spot for horses, dogs and sheep. Pet sheep, dogs and hens were regular invaders of the house, and several pet sheep at Ellisland are mentioned.

The milk products of cattle were of great importance, as in both the old and new farming they generated cash income. In Burns' day the native cattle of Ayrshire — probably not unlike the black Galloway of today, but with a variety of colours — was crossed with Shorthorn stock from the north of England, some of which came via Berwickshire. The result was the Cunningham or, as it became known, the Ayrshire breed. This became the foundation of a dairy industry that has continued unbroken. The milk cows Burns had at Ellisland may have been Ayrshires, for he recommended the breed to the minister of Dunscore.

Butter and cheese making were in the hands of the *guidwife*. On moving to Ellisland, Burns had a problem because, bonny though she may have been, Jean Armour was not a farmer's daughter. She stayed on at Mossgeil to learn the important skills from Agnes, Burns' mother. One of the improvements in the new farm buildings then going up was a separate *milk-hoose* or dairy. The butter was made by allowing the cream to *settle* or rise in broad shallow dishes, skimming it off and churning it. A *kebbuck* or cheese was made by curdling the milk with *yirnin* or rennet, breaking the *crudes* or

above: Northern Isles sheep, about 1845.
These were only one among various native breeds
of sheep in northern Britain in the late eighteenth
century, including the old Dun-face and White-
face. With the exception of the Soay sheep, of very
remote origin, they were fine-wooled. They were
also small and hardy, and able to withstand the
rigours of winter on lean rations.   William Shiels,
NMS

left: Dairy equipment such as this *plowt kirn* or
churn, *milkin luggie* and milk *boyne* were usually
made of finer hardwoods so that they did not taint
the milk. Although it is possible to do without,
the milking stool made a tedious job much easier.
The shallow horn skimmer is for taking off
cream.   NMS

right: Milking was done early morning and late afternoon, and together with dairywork was a constant burden on the womenfolk.   Walter Geikie, National Gallery of Scotland

Soutar Johnnie, the shoemaker in *Tam o' Shanter,* was probably Johnnie Davidson, who lived at Glenfoot at Arlochan, near Shanter, and later at Kirkoswald. Although some people made their own shoes, usually crude *straucht shune* or 'straight-shoes', that would fit either foot, the shoemaker's was one of the older domestic trades, along with those of the tailor and the weaver.   SEA, NMS

curds into an even consistency, and then pressing that in a *chissit* or cheese vat. The Dunlop cheese of Ayrshire became famous for its pleasant mild taste.

Skills, occupations and people

The new farming brought a shift in the balance of the population, because it demanded the expansion of specialized trades on which it depended to service it. With the new metalled roads, horses needed more shoeing. More plough coulters and socks had to be relaid. The dished and spoked wheels and the handsome new carts required the skills of joiner and smith alike. Horse *graith* or harness was increasingly made by the saddler, and not home-made. The corn mills were being upgraded, often with new-style cast components from the Carron Ironworks, near Falkirk, and this required the skills of the millwright. Where before people had built their own and their neighbours' houses, a whole new rural building trade with masons, joiners, slaters and thatchers was coming into being. With general economic growth there was more work for the traditional domestic craftsmen such as the tailor and *souter* or shoemaker. These are the people who crowded the fireside in *Tam o'Shanter.*

The craft industries which were now expanding round the old country towns were also adding another dimension. Arkwright had discovered how to spin yarn mechanically, but it still had to be woven by hand. This gap was filled by a spectacular growth in handloom *wabsters* or weavers. Like miners, they came to form a distinct subculture, neither entirely urban nor rural. Burns noted them in 'The Holy Fair':

> An' there, a batch o' wabster lads,
> Blackguardin frae Kilmarnock
> For fun this day.
>
> *The Holy Fair*

Burns had a marvellous eye for scenes of roistering merriment, the 'merry core o' randie gangeral bodies' that met at Poosie-Nansie's in Mauchline. He drew a sympathetic caricature of the considerable floating population of unemployed, tinkers, discharged soldiers and their followers, and general foot-loose characters. This was partly an inheritance from the past, when

Market day at the Inn. There were no shops in the country, only markets. What small things people did not buy from the *chapmen* or *cadgers* – travelling salesmen, or get made by local tradesmen, or bring back from an occasional visit to a town, they would get at the markets. Drinking was part of social ritual, and either sealing or celebrating a bargain at the market called for refreshment.   Alexander Carse, National Gallery of Scotland

homelessness and beggary were a serious and insoluble problem. Before farming improvement, there was much less casual employment available to absorb this population.

Although the life of the 'jolly beggars' must have been far from jolly once the spree was done, they were nevertheless people who could shift for themselves. For the 'deserving poor' — the old and the infirm — there was respectable beggary, or *aliment* or assistance from the parish funds.

It was to the people on the lowest level of the settled population that Burns' heart went out. Cottars were married people with children to feed and clothe. They might be hired farm servants, but there the openings were limited, for in the Ayrshire of the time, farmers relied mostly on their own sons and daughters, or unmarried servants, often the cottars' young folk.

Mossgeil was near Mauchline, the home of Jean Armour, and the place of those opposites, the parish kirk and Poosie Nancy's Inn. From the late seventeenth century there was a growth of villages, some of them 'burghs of barony', or the centres of a 'regality' as was Mauchline. With improvement, these places came into their own, providing useful local markets and centres for tradesmen, such as Jean's father, who was a mason.   Anon, National Gallery of Scotland

opposite: Wheelwrighting was a new skill. To old tools – the saw, dividers, draw knife and brace were added newer tools – the spoke gauge and *bruzz* or two-sided chisel for mortices.   NMS

This noble image conveys how 'The Cotter's Saturday Night' caught the imagination of succeeding generations. The image got into the bloodstream of nineteenth-century radical politics. Keir Hardie, the founder of the Labour Party, was a fervent Burnsian.   William Miller after John Faed, City of Edinburgh Art Centre

However, besides part-time work on the farms, the cottars' livelihood was mostly whatever labouring jobs they could pick up:

A cotter howckin in a sheugh,	digging a ditch
Wi' dirty stanes biggin a dyke,	building
Baring a quarry, an' sic like,	clearing
Himsel, a wife, he thus sustains,	
A smytrie o' wee duddie weans,	a collection of ragged little kids
An' nought but his han'-darg, to keep	hand's work
Them right an' tight in thack an' rape.	thatch and rope

The Twa Dogs

opposite: Burns writes to John Murdoch, his old teacher, confessing that although he is not idle, nor is he a 'pushing, active fellow', but one who delights to 'study men, their manners, and their ways'. Although the letter is a conventional flourish of style, Burns was perhaps closer to the mark than he knew.   Trustees of the National Library of Scotland

Burns drew a picture of decent people with a canny ability to make the best of an unenviable place in life, and even enjoy themselves. But in 'The Cotter's Saturday Night' he did something that astonished his contemporaries. He ascribed to the landless and near destitute labourers of the Ayrshire countryside the virtues of literacy, intelligence and a simple and sufficient social and spiritual grace, and asserted that it was 'from scenes like these old Scotia's grandeur springs'. Although an idealized portrait it is a brilliant icon of the dignity of labour.

has been the "result" of all the pains of an indulgent father, and a masterly teacher; and I wish I could gratify your curiosity with such a recital as you would ~~what you was~~ be pleased with; but that is what I am afraid will not be the case. I have, indeed, kept pretty clear of vicious habits; & in this respect, I hope, my conduct ~~will would~~ not disgrace the education I have gotten: but as a man of the world, I am most miserably defficient — One would have thought that, bred as I have been under a father who has figured pretty well as un homme des affaires, I might have been, what the world calls, a pushing, active fellow; but, to tell you the truth, Sir, there is hardly any thing more my reverse — I seem to be one sent into the world, to see, and observe; and I very easily compound with the knave who tricks me of my money, if there be any thing original about him which shews me human nature in a different light from any thing I have seen before. In short, the joy of my heart is to study men, their manners, and their ways; and for this darling subject, I chearfully sacrifice every other consideration: I am quite insolent about those great concerns that set the bustling, busy sons of care agog; and if I have to answer the present hour, I am very easy with regard to any thing farther. E

MAN'S DOMINION

If the land Burns knew was changing, it was to change even more after his death. Effective sub-soil drainage would realize the full fruits of improvement, and the familiar *rigs* would be levelled, opening the road for horse-worked field machinery. The orderly garden of the Agricultural Revolution would triumph over the ramshackle untidiness of the old landscape and its buildings. But it is this same land which was the constant backdrop to Burns' poetry, and not unlike the women of his experience, it was the source of inspiration and heartbreak. Yet it was never a land of disembodied romance. It was cropped and stocked and peopled. The tradition of animal poems in which Burns delighted all integrate their subjects into the human framework. Speaking to the little wild beast he has disturbed, Burns is

> … truly sorry Man's dominion
> Has broken Nature's social union,
> An' justifies that ill opinion,
>   Which makes thee startle,
> At me, thy poor, earth-born companion,
>   An' fellow-mortal!
>
> *To a Mouse*

It is an imperfect but single world. Of the imperfections Burns was well aware, including his own share in them. Yet whatever the difficulties he got into with his love affairs, the worries he had over money, the scrapes he had with civil or religious authority, the burden of ill-health about which he rarely complained, he never lost his sense of those wonderful moments when he was at one with the land that had bred him:

> Upon a simmer Sunday morn,
>   When Nature's face is fair,
> I walked forth to view the corn,
>   An' snuff the caller air: fresh
> The rising sun, owre Galston muirs,
>   Wi' glorious light was glintin;
> The hares were hirplan down the furs,   limping; furrows
>   The lav'rocks they were chantin
>   Fu' sweet that day.
>
> *The Holy Fair*

opposite: Combining corn at Galston, Ayrshire, 1989. Here the grain is processed straight from the field. To achieve the same result in Burns' day, the crop had to be sheared, bound into sheaves, and then stooked. When dry and further ripened, it was led into the cornyard, and built into rucks which were then roped and thatched, where it would *win* further. The stacks were dismantled one by one through the following months and the grain thrashed from the straw with the flail, sheaf by sheaf. Only when the chaff had been winnowed from the threshed grain was the same stage reached as shown here.   NMS

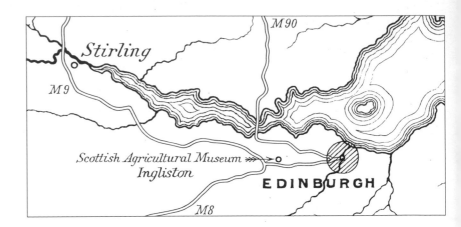

The agricultural and rural life collections of the National Museums of Scotland are housed at the Scottish Agricultural Museum which is open from April to September each year.

ACKNOWLEDGEMENTS

We would like to thank the following for help with the illustrations for this book: Lorna Ewan; James Hunter, Dick Institute, Kilmarnock; Tom McIlwraith, Past President Burns Federation; Dr Rosalind Marshall, Scottish National Portrait Gallery. We are also grateful to Dr Alexander Fenton for commenting on the text. Pictures credited 'SEA' are from the NMS Scottish Ethnological Archive.

opposite: The banks of the Nith, where Burns wrote *Tam o' Shanter.*   NMS

outside back cover: Galston. The only thing Burns would recognize now is the crop of corn. Elizabeth Robertson